CW00497211

TIM CRESSWELL

Tim Cresswell is a geographer and poet. He is the author or editor of over a dozen books on the themes of place and mobility. His most recent title is *Maxwell Street: Writing and Thinking Place* (University of Chicago Press, 2019). His poems are widely published on both sides of the Atlantic, including in *The Rialto*, *Poetry Wales*, *Magma*, *The Moth*, *LemonHound* and *Salamander*. His two previous collections of poetry, *Soil* (2013) and *Fence* (2015), were published by Penned in the Margins. He co-edits the interdisciplinary journal *GeoHumanities* and is the first Visiting Professor at the Centre for Place Writing at Manchester Metropolitan University. Tim lives and works in Edinburgh where he is Ogilvie Professor of Geography at the University of Edinburgh.

ALSO BY TIM CRESSWELL

POETRY

Fence (Penned in the Margins, 2015)
Soil (Penned in the Margins, 2013)

CRITICAL WORKS

Maxwell Street: Writing and Thinking Place (University of Chicago, 2019)
Place: An Introduction (Blackwell, 2014)
Geographic Thought: A Critical Introduction (Blackwell, 2013)
On the Move: Mobility in the Modern Western World (Routledge, 2006)
The Tramp in America (Reaktion, 2001)
In Place/Out of Place: Geography, Ideology and Transgression (University of Minnesota, 1996)

Plastiglomerate

Tim Cresswell

Penned in the Margins

LONDON

PUBLISHED BY PENNED IN THE MARGINS
Toynbee Studios, 28 Commercial Street, London E1 6AB
www.pennedinthemargins.co.uk

All rights reserved
© Tim Cresswell 2020

The right of Tim Cresswell to be identified as the author of this work has been asserted by him in accordance with Section 77 of the Copyright, Designs and Patent Act 1988.

This book is in copyright. Subject to statutory exception and to provisions of relevant collective licensing agreements, no reproduction of any part may take place without the written permission of Penned in the Margins.

First published 2020

Printed in the United Kingdom by CPI Group (UK) Ltd

ISBN
978-1-908058-76-8

This book is sold subject to the condition that it shall not, by way of trade or otherwise, be lent, re-sold, hired out, or otherwise circulated without the publisher's prior consent in any form of binding or cover other than that in which it is published and without a similar condition including this condition being imposed on the subsequent purchaser.

CONTENTS

ACKNOWLEDGEMENTS

I am grateful to the editors of the following publications where some of these poems have appeared: *Clarion, The Clearing, Frogmore Papers, Kudzu House Quarterly, Reliquiae, Sandy River Review, Soundings Review, Spiral Orb, Stare's Nest, Stockholm Review of Literature, Tears in the Fence* and *Transfers.*

I have benefitted from the input of many people while working on these poems. Thanks particularly to Suzanne Buffam, Philip Gross, Andrew Motion, Srikanth Reddy, Jo Shapcott and Karen Solie as well as all the participants at the Banff Writing Studio, Faber Academy, and Arvon workshops that I attended. I have benefitted enormously from the critical input of members of Tom Daley's Monday poetry group in Cambridge, Massachusetts and Jim Finnegan's Brickwalk Poetry group in West Hartford, Connecticut. Thank you to Julia Barton for supplying the plastiglomerate for the cover of this book.

Plasti-glom-erate

Rialto Beach

first   the carnage
silverwhite trees   wave-toppled and gale-gusted

strewn along the beach   sun-bleached dinosaur bones
next   the balancing acts of stonestacks   a lone conifer

clings to invisibly thin soil   knots of bull kelp
fetid and fly-ridden   anemones in rock pools

beckon me to finger   their sticky tongues
a bald eagle circles   hounded by ravens

we keep on holding on   hoping for the best
settling   for something like circumstance

in the black basaltic sand   mingled with milk-white
and amber agates   smooth ground sea-glass jewels

among the fallen trunks   chunks of rusted ship-iron

scattered fishing floats   a plastic buoy

a walker   tells me of the wreck   the Chilean ship
the marker 'round the headland   for the drowned

Plastiglomerate

All-purpose cobalt polypropylene rope
sluiced overboard by Tom   ten days out

with the tuna fleet   missing Meg
and local radio   ground down shells

clams and cowries   orange-lipped lava flows
the toothbrush Esme shared before she spat

foam into the breakers   bleached
pummeled coral   thin plastic forks

from when you barbequed
hot dogs on driftwood fire   fucked

on the foreshore   shared photos of your children
abraded black basalt   green beads of olivine

the chewed blue cap of the one-buck biro
that leaked in the heat of my pocket

forest fire   twenty-three percent contained
fifty-five gallon polydrum   ready for rain

Nest Site Fidelity

Ospreys are back from
wintering in Aruba—

skywriting *M*s and *W*s
then plummeting

talons slicing brackish water
rising with a prize menhaden.

They return each March
to the knotted nest they knit—

wayfinding magnetic fields with
star charts and landmarks

Haiti   Havana   Florida Keys and Everglades
Jacksonville   Savannah

then this tumbledown

stick-and-spit summer-home

on the platform Bill built
here on the Chesapeake.

Me? I've flown in from Wales
where the Emperor of Japan once visited

to spy the rare red kites
I saw on my daily drive

cresting the rise
above the Irish Sea.

Scale

Banks collapse, beds need changing, a girl
in a veil prepares to kill herself.
Iceland's for sale. Ironing piles up.
Food festers in the freezer's broken trays.
Fevers swarm in vectors, thick black arrows
thrust across the page, viruses spreading
from who knows where. Words, misheard, fall
pointlessly between us, on washcloths,
recipes, pot plants. In the troposphere,
atoms swerve, winds diverge, a lifting
air mass gifts a polar vortex.
We're forecasting snow and record lows.

Car Plant

There's a Morris Minor in the yard,
Morris Green, a sunny yellow sticker
from the seventies shouting
NO THANK YOU to nuclear power.
The car's name is *Bumble*, or *Bramble*.
Sparrows nest on the wing mirror;
green moss grows on the window frames;
there are holes in the floor beneath
the mats. Flakes of rust
stain soil ochre. Old engine oil,
or brake fluid, spreads shadows.
Buddleia clings to wooden trim,
drawing a rabble of Peacocks, Admirals,
Swallowtails. *Bumble*, or *Bramble*,
hosts xylem and phloem sucking
nutrients through tyres,
along axles, through the rear shocks
and brake pads, into the chassis.
Roots bore down from wheel wells,

telling the soil-dwellers stories of day-trips
to the seaside and the passwords
of local girls who knew the secret
places where the keys were kept.

Haul Out

Finding no ice in the Chukchi Sea
thirty-five thousand walruses
hauled out and huddled
near Point Lay, Alaska
with its Cold War radar array
and two hundred and forty-seven souls
six of whom are 'Latino or Hispanic'
which makes me wonder
how they got there and why.
The walruses bellow, snore
walk with their teeth
take turns diving deep
whiskering for mollusks
whose numbers are declining
like the Kuukpaagruk
who dwindled down to two
thanks to Yankee whaling ships
disease and decimation.
Here come the helicopters

and cameramen, clamouring
for coverage, risking a stampede.
The walruses hunker down
waiting for a freeze-up.

In a Station of the Underground

after Ezra Pound

Out of the underground
   hot from the
      crowds—

streetlights mirrored
   in the storefront
      rainbow in the oil

rain snaking down
   a Perspex screen.
      Don't speak to me

of rivers, salmon
   leaping falls
      a forest's bluebell haze

or granite scoured

by ice.
　A woman weaves

against the flow
　heading home
　　her face a full moon.

The Two Magicians

after Child Ballad 44: 'The Twa Magicians'

The lady stands in her bower door,
    As straight as willow wand;
The blacksmith stood a little forebye,
    Wi hammer in his hand.

        'Weel may ye dress ye, lady fair,
            Into your robes o red;
        Before the morn at this same time,
            I'll gain your maidenhead.'
'

I'd rather I were dead and gone,
    And my body laid in grave,
Ere a rusty stock o coal-black smith
    My maidenhead shoud have.'

then she became a turtle dove
    purring in the linden

and he became another
　　　　strutting　　swelling
　　whistling his wings　　rising
　　　　to pair with her

꙰

then a hare
　　on the hill
standing　　twitching evening air
　　mad as anything
　　　　ready to box

　　and he became a greyhound
　　　　sprung from his master's side
　　　　　　ears swept
　　　　　　　　sinews straining

꙰

then a gay grey mare
　　　standing in the slack

　　　　　and he became a saddle
　　　　　　　and sat upon her back

　　　　　　　　ॐ

then queen bee
　　　attended by armies
kissing clover
　　　massing in meadows

　　　　　he became monoculture　　　mites
　　　　　　　the absence of orchids
　　　　　　　　　trucks of hives on the highway
　　　　　　　　　　fruit farms in the valley

　　　　　　　ॐ

then she became peregrine
  swooping from her scrape
stationary
  then falling
     stone
  a shooting star you thought you saw
lightning
     arrow

    he became leather hood for manning
       strong jesses
          bells
    he became endocrine disruptor
    persistent organic pollutant
       with lipophilic properties
    he became hydrophobic
      but soluble in fats
    colourless
      tasteless
        odourless   mostly
    he became crystalline
      travelled under many names

Anofex Cezarex Chlorophenothane Clofenotane
Dicophane Dinocide Gesarol Guesapon
Guesarol Gyron Ixodex
Neocid Neocidol
and Zerdane

ક્ર

then the moor
    purple blush of heather
        paths tracing habits
            and heft of sheep
the common land
    bound only
        by the circling sky

        and he became
                An Act for the better Cultivation,
                Improvement, and Regulation of the
                Common Arable Fields, Wastes and
                Commons of Pasture in this Kingdom

and said

> ... all the tillage or arable lands lying in the said open or common fields shall be ordered, fenced, cultivated and improved in such manner by the respective occupiers thereof, and shall be kept, ordered and continued in such course of husbandry, and be cultivated under such rules, regulations and restrictions...

> Etc.

❧

then grey whale with her enormous hunger   sieving oceans
    luminous lives
        jiggering
            suspended

and he became a multitude

duct tape

duct tape

electrical tape

fabric – miscellaneous

fabric – miscellaneous

fabric – miscellaneous

fabric – miscellaneous

fabric – miscellaneous

fabric – sock

fabric – sweat-pant leg

fabric – towel

fabric – towel

fishing line

golf ball

nylon braided rope

plastic – red plastic cylinder

plastic – black fragments

plastic – Capri-Sun juice pack

plastic – grocery bag

plastic – Misc bag material

plastic – Misc bag material

plastic – Misc bag material

plastic – Misc bag material

plastic – Misc bag material
plastic – Misc bag material
plastic – Misc bag material
plastic – Misc bag material
plastic – Misc bag material

৵

then air
       wind carousing the barley stalks
lifting   then
      sinking
        tickling the aspen
  shepherding leaves against the hedge that lines the
      hollow way
  spreading
        spiralling in
          the sky circling the fens
     in summer
   lifting   sinking   circling   and she was called
sirroco   mistral   bora   chinook   zonda

willy-willy and williwaw   lifting   sinking   circling
and he became phyloplankton and zooplankton
        terrestrial flora   dead   sedimented   crushed
        mixed with mud   heated and pressed   folded
and he became kerogen   shales   gaseous
        hydrocarbons ethane propane butane
        pentane methane sour gas tight gas
        shale gas sweet gas
and he became anthrocite dirty lignite bitumous
        coal fat coal forge coal flame coal
he became paraffin aromatic hydrocarbons
        crude oil bitumen kerosene gasoline
        petroluem diesel
he became internal combustion
he became one part carbon two parts oxygen
        nitrous oxide ozone
and he became auk clyde fulmar shearwater
        gannet blane kittiwake nelson
        forties buzzard golden eagle
and he became argyle duncan elgin-franklin
        erskine merganser pierce arbroath
        montrose lomond everest britannia
        alba buchan andrew balmoral

        gryphon   thelma   he became   piper alpha
        ocean ranger   deepwater horizon
he became   Abathasca
he became night shifts   fly-ins from St John's
        three weeks on   one week off   an unwanted
        hand on a thigh in Fort McMurray
he became five-mile tailbacks on the interstate   a lone man
        in a car   tuning into   Fox FM   *we ride through*
        the mansions of glory in suicide machines
he became   GMC Sierras   Chevy Silvarados
        Toyota Tundras   Nissan Frontiers
        sprung from cages out on highway nine
he became   rock fractured by high pressure liquid

                          ⚘

then she became earth tremors in the kitchen in Norman
Oklahoma   coffee   Patsy on the radio   snowdrops in January
in Kew   *crazy*   iguanas   frozen   falling from trees   boiled
bats in Sydney   Harvey howling like a freight   *crazy for*
trying   and *crazy for crying*   iceberg   twice   the   size   of

Luxembourg *worry* thirty-five thousand walruses hauled out in Point Ley Alaska *why do I let myself worry* Irma manitees swimming round a drowned Winnibago ninety nine percent of green turtles born female *wondering what in the world did I do*

A Theory of Migration

Push. The slowing down of everything
but you. Like when the driver hits the brakes
and you're slo-mo through the windscreen.

Pull. The earthly tug of gravity
on space debris. The fizzing blue light
that zaps the bugs by a turquoise sea.

Heathrow

My Somali taxi driver listens
to love pop on Fox FM, driving

with one hand and abandon up the ramp
in the outside lane, cutting in at the last

second to make the M4 — almost empty
in the early hours. I guess he's from the tower

blocks south of the shops that sell SIM cards
with calls to Africa. 3p a minute.

I'm off to Chicago, passport in pocket,
briefcase stuffed with paper, laptop, charger,

staring at my phone. The Amazon's on fire.
There's plastic in the blood cells of a blue mussel.

A woman in Lagos wants to send me a million dollars.
There's one weird trick I need to know.

Legend

Do I need to know which way is north?
I know the light is different there
    that water freezes blue,
    white, pink.
That snow piles up and remains
    as reminders
    into April, dirtied with diesel
    and particulates.

Do I need to know the scale of lives?
I know we can be longitudes apart, over
    oceans and rifts where continents
    submerge and drift.
That we can be close enough
    to breathe each other's exhalations
    and sometimes distant scents
    of cities both familiar and strange.

Do I need to know what all these symbols mean?

I know the places that I love and hate,
 what's high and breathless,
 beautiful and desolate.
That we have troughs and summits,
 the rise and fall of U-shaped valleys,
 contours once tightly packed
 now hardly there at all.

Do I need to know who made this map?
I know there's strangers on a train
 with airport novels
 and bitten nails.
That there's good coffee to be had—
 music's infinite array of sounds
 the truth of hands
 and trinkets to be found and lost
 foxes at midnight
 scavenging in bins
 and herons with embracing wings
 waiting to take flight
 while out at sea and in the unmarked space
 there be monsters—things unknown
 luminescent swimmers sinking down

lantern bright—leaving no trace
and north, far north of you or I,
on some icy island beach
are whalebones, brittle, bleached
under magnetic Arctic skies.

In Brookline, Massachusetts, I learn a new route

a greenhorn with yellow running shoes.
Everything's different, or the same
just five degrees askew.

I avoid wild turkeys on Blake Street.
Cicadas bow their seventeenth-year symphony,
an off-beat Philip Glass diminuendo

thrumming all my lesser bones.
Markets end in the red zone thanks to sluggish growth
in Europe and air strikes in Syria.

Carol said turkeys remember a hundred human faces.
Orthodox neighbours don wigs to stand around the reservoir
tossing in their sins. Chinese women sit beside the water

dressed in sweaters, winter coats and baseball caps.
They fish for fish I do not know:

bullheads, pickerel, pumpkinseed.

This morning my son, Sam,
wore a lumberjack shirt and blue nail polish.
His new friends call him Alice.

Newfoundland

A woman in St John's
says she likes my scarf
in a chock-full bookshop
called Afterwords. It's a thing
to get shit-faced on George Street.
A band called Quaker Parents
plays at The Ship. People sound Irish.
The cod have long gone
but now there's oil.
There is dark rock, thick fog
and thin soil. To belong
for a while, out-of-towners
can drink screech, kiss a cod
and repeat after them,
"deed I is me ol' cock,
and long may your big jib draw."

I can't count the times I've flown over
sleepless, beat, following

the blue plane on the seat back's moving map.
Thirty-one thousand feet,
counting down the miles.

In December, zero degrees,
unlikely clear blue sky, I'm driven
in a battered Subaru past the jellybean homes
to Cape Spear. Sea churns froth against
the rocks; a hollow-eyed
bunker stares down the grey Atlantic;
two lighthouses, one old,
one new. My companion,
a lanky local with scraggly beard,
recites the names of capes and coves.
Portugal Cove. Cuckold Cove.
Quidi Vidi. Cape Despair.
He speaks of moose on Water Street,
synclines, anticlines, short-tailed weasels,
iron mines, the comforting absence of bears.
Back past Signal Hill, across the barrens
and into the boreal forest, above,
four, five, six white trails of planes
flying straight, in loose formation,
following an aerial highway west.

Beached

Fishermen find it at dawn, spread across the tidal zone.

Workers from the shopping mall take lunch breaks on the seawall.

Agnes Connolly starts slipping global warming into casual conversation.

Townsfolk take turns posing with it, gathered in family groups; some statue-proud, others waving two-fingered peace signs.

Down at the dealership, sales of pick-up trucks pick up.

Parents, half-cut on Shiraz, read stories to their kids.

Children circle, dare to touch, learn its heart is the size of a Volkswagen Bug and its tongue three tons. John Connor lets it be known in furtive whispers that it has a ten-foot cock.

Three members of the Rotary Club resign.
Old man Jude at number 79 complains of a sweet sick smell.

Some see it as a sign. Wailing is heard from the Methodist church – the minister intones a litany of species.

Its body begins to bloat, blow up; a balloon about to float away.

Clouds of stink waft landward. The librarian carefully catalogues references. Rancid bacon grease, rotten fish, eggs, cat pee, concentrated cow-farts. It might become bearable, she thinks, with the right words.

Jim O'Leary, while making love to his mistress, imagines his semen forming clouds of milky nebula in the briny dark.

The Wolverines end a season-long losing streak in the dying moments.

There are rumors of exploding whales at Foodland and at the bowling alley. Methane accumulates. #explodingwhale appears on Twitter. Updates on hasthewhaleexplodedyet. com. In dreams, whale bits rain down, whale-juice coats the town in an oily blubber-sheen.

Mothers avoids the seafront, take roundabout routes to the movies, peep through curtains. Hotels stay vacant.

It is generally agreed that something should be done.

Max Jenkins goes the wrong way home. Finds himself, key in hand, at number 41.

Young Bill Shaughnessy, who never speaks to his parents, can't stop talking. Says he saw two white tigers strolling the lapping waves under a full moon, rubbing their flanks against its grooved throat.

Kate McConnell, a lifelong omnivore, finds herself disgusted by bacon, starts hoarding kale chips.

Lives fill with whale-chat or silence. It expands. The stink creeps down thoroughfares, along back alleys, curls around cats, sceps through grates and gaps into ductwork, sweaters, high performance fleece. Masks are issued.

Dick Declan, while watching birds, sees a Labrador Duck, long thought extinct.

An artist arrives from Whitehorse to film the slow decay, how the blubber sags and falls away. Another, from Yellowknife, requests secret mementos to arrange along the beach. Miss Levine's Grade Five art class suggests five hot air balloons, decorated, whale-shaped, launching from the beach at dusk.

Plans are hatched. Dynamite. It could be burned or buried where it lay. Tow it out to sea, laden with old train wheels.

Adam Smith, a local anthropologist dropping all pretence to objectivity, declares the whale divine (in the tradition of Austro-Asiatic cultures) and suggests a funeral. Incense would be burned, libations poured, a sand temple constructed.

Museum men arrive, fillet with flensing knives. Strips of blubber, meat and muscle pile on the sand. They are greedy for bones—a skeleton in the central hall! A place to meet.

Suddenly fearing its absence, the Mayor suggests a waterfront display—something for kids and tourists. Educational but fun. Townsfolk cluster in circles, sing shanties. Hold a candlelit vigil.

When the work is done and all that's left are bones, they ship them out on Mack trucks.

Fugitive Pigments

here come the Mexicans with their leaf-blowers
and surgical masks   moving the auburns
chestnuts and russets from here to there

marshalling pigments that shift shades
or fade—   memories turn sepia   vegetables grey
become less nutritious   look!

the trees are dying for us   again!
green rusting   edging yellow   orangebrick
Bordeaux red   burnished bronze   windblown—

verging along the curbs   gathering against
garage doors   the municipality has issued a decree
no shade to cross property boundaries

especially ones that quickly lose depth upon
exposure to light   use colour-cleaning devices
where feasible   to capture remains

of cadmium yellow   ultramarine
before intensities are abraded
and dispersed   becoming

subdued browns and lesser greens   winterworn
with persistent blackened piles of snow
battalions of cumulus clouds   flat   grey-

bottomed   the trees are ending again
dust off travelling booms
telescopic chutes   rotary stackers

impose strict   slow   speed limits   in accordance
with recognised and generally accepted methods
as determined by the Department

Tremor

Alice has painted nails
fingers splayed   fingers down
each one a flag that means
something to her   this one
transgender   that one asexual
and one rainbow even I understand.
She's off to march
on the common   stopping
traffic   shouting
I can't breath.
I should go too
but I want to tell her about
sitting on the fully automatic
deodorising toilet
how the room swayed
a slow sway, and swayed
again, slowly in waves.
Returning to bed
in the tiny Tokyo

room. I slid
the screen to peep
out at the street
across to the offices where
white-shirted salarymen
also took a peek
then returned to their desks
as if nothing mattered.

Dendrochronology

here the factories came alive
    the era of trains
    ozone and petroleum
Hiroshima
    contrails across the sky
the week in Magaluf
    humvees
    the end of elms
    the rise of rhododendrons
    recession and recovery

a summer when the rain held off
    an unexpected frost
a year that lingered
    a year that flew
steady rise of isotopes
    fate of honey bees

here too

rain dripped
through the canopy
on lovers taking shelter
in each other

in its bark
a mark
once freshly knifed
and full of love and sap
now gnarled and whorled
but visible

look

Spoil

marks the
old deep mines
behind the terraces
and gorse   too smooth
for wind-battered ice-pucked
land   birthed from under   where
blackened hands excavated anthracite
useless slagslurrysandstoneshale landslip
tailing dung heaps of the origins of industry
dumped by some incontinent Goliath after the
good stuff's been extracted   ruin and plunder
kilowatts created   steam trains   ships sent
on their way   central heating   crimson
sunsets   rain stripped trees   rivers
break their banks   snowdrops
in December   melting ice
lost birds   spiral past
the mine museum
picnic spots
oh my god
the view

Erratic

Sitting there for all the world
as if you own the place.
An error. Bluestone among
chalk. Wanderer. Heretic.
Stowaway, suspended as the world
warmed leaving you upended—
culture-shocked and supersized.
Your crystals milky,
opaque, glittering.
Brimming with another kind of
winter. Fire and freeze. Ten
thousand years of exile—
now part of the scenery.

Friendly Floatees / *Tripadvisor*

Sometime on 10th January 1992, the container ship *Evergreen Ever Laurel*, en route between Hong Kong and Tacoma, Washington, encountered stormy seas. Containers were washed overboard. In one there were almost 29,000 plastic bath toys known as 'friendly floatees'— green frogs, red beavers, blue turtles and yellow ducks. The toys were manufactured in the Pearl River Delta in China for Kiddie Products of Avon, Massachusetts. Somehow the container split open and the toys continued their voyages.

At Sitka, Alaska's second annual beachcomber fair, Dean Orbison and his son Tyler exhibited 111 plastic frogs, turtles, beavers and ducks they had found between 1993 and 2004.

> *take*
>
> *some time*
>   *enjoy   the beach*
>
> *the stream   was full*
>   *of salmon   very cool*

counted
      17 bald eagles
      bears are sighted
            every year

loaded
with totem poles
      Sitka history

          at low tide   rocks
                  with barnacles galore
some blueish shells   singing ravens
      super-sour huckleberries
          beach
          forest
          totems
      humpback whales   hanging out

kids can have fun
    in tide pools

Tlinket culture
      informative film

you might get to see

           a crane

           or two

the best of
    all worlds

On the Queen Charlotte Islands, in British Columbia, Guthrie Schweers found two blue turtles and four green frogs. This was 1995.

what can I say?

the drive there
    is spectacular

loved walking
    on the beach   saw unknown
sea creatures   a beautiful place
    to unplug   listen
      eagles galore
        many shore birds

go hunting
    for agates

beaches are   *endless*
razor clams   *plentiful*

very Zen!

if you wander
far enough
you will also find
fossils
a backdrop
of rainforest

reconnecting
with yourself

waves and storms
reveal more

lots of neat rock
it doesn't get

much better

In July 2003, Bethe Hagens and Wayne Welton spotted a yellow duck at the west end of Gooch's Beach in Maine.

get there by nine
        to find
        a parking spot

our all-time favorite

    there are waves
        not huge and scary
a great place
    to bodyboard

all the beauty
    of a bigger beach

        sand dollars at low tide
            some live crabs
        kids enjoying   hotdogs

I wasn't sure
    what kind of sunset

I would get

sidewalks   nice and wide

collecting   sea glass
is a fun activity
tranquility is best

water
getting
closer
and closer
horizon
goes on
forever

Sonali Naik, an English barrister, found a green frog on Uig Sands while holidaying in the Isle of Lewis, in the Hebrides, in 2006.

sometimes

when you set out
on a journey

you wonder

        single-track roads
        light blue sea
        miles of white sand
        no warning signs
            whatsoever

at low tide
    you can wander
        almost
the whole way
           across

    immaculately clean

    plenty of space

free parking
great for flying kites

a must visit

pink/orange rock
        of the type
          I've seen in Iona

hold hands and beachcomb

kids cried
and were soaked

adults loved it

       a long way
from anywhere   that
         end of the world feel

hard to find
        directions
          but worth it

when you get there

What I said was

the biomass of squid exceeds that of humans.
There are new islands forming off Iceland
and worms a mile deep with teeth.

What I meant to say was

    you can still sit
    in amongst the bluebell
    haze of woods in May,
    your history erased. An early bee,
    sluggish in the shadows,
    the breeze laden
    and feel your body. Full. Then.

In the Natural History Museum

Among Mastadons
and granite
skeletons
and meteorites
the cabinet
of hummingbirds;

not the one on a
wooden porch
south of the Mason-
Dixon Line flying
at the feeder
invisible wings
bumblebee impossible
tonguing sugar-water.

Harvest

The mice returned, betrayed themselves
with droppings on the table.

As autumn bedded in they sought out kitchen warmth;
ran along countertops and shelves.

We didn't like killing them, finding their
small stiff bodies disturbing, the wire

in the flesh and blood pressed out
of heads through noses drying on the trap.

(In our defence we'd read that you have to move
a mouse a mile from your house to stop it
from returning).

With winter just around the corner, the acre
we owned was stripped for silage

by Idris down the hill with his
John Deer and baler. He gave the field

a buzz-cut, turning tall grass into
picture-book bales.

It was an unplanned massacre
of mice, minced and quartered

in the stubble. Paddy bounded out
half mad, chucking back the heads and tails.

Fulgurites

The most fleeting things are forever if you know where to look.
I Googled them—spotted one on eBay.
'Lightning power in the palm of your hand!'
Mostly, they   scatter the Sahara suggesting
                              the presence
                          of rain or
                        Zeus or
                        Thor
                    the opposite of sedimentary—
                  petrified milliseconds
                formed in
              a flash
                when quick high volts struck
              shifting sands
            quartz and
          silicon
        fused into glass of a sort—
        tubular molds in
      the shape of
      less than
    a blink
    of an
    eye
such as the stegosaur's. The only witness to the fireworks
and thunder when the oldest ones were formed.

Flaws

cracks appear
　through the Polyfilla
wallpaper
　in the bathroom
　　curls—peels
　　　away
mildew blackens grout
red wine stains
　the grain
　　of the kitchen
table

bindweed blooms white tubas in the yard
lichen sketches goldgreen archipelagos
on breeze blocks

there's a flaw in our Persian rug—
sudden
　tick of turquoise

in the woof
   and weft
      of red
put there by the weaver
not wanting
   to presume
      paradise

Footfall

after Jean Sprackland

That strip of brass
that lines the edge of the step

bears the weight of us
through days and decades—

the streams of feet—the multitude
who cannot help but step on you—

as we exit Marble Arch to nightclub,
Primark, penthouse, office block.

All that varies is the footfall
at rush hour, or the fag-end of the day;

the volume of traffic, the metre
and velocity of clamouring.

Stubborn. Resilient.
Consider a mountain.

Blues for Lost Birds

Pigeons echo, from the roof and the linden trees
Pigeons echo from the roof, from the linden trees

Swallows swerve and chase down flies
Swallows from Johannesburg, tracing the sky

Yellow-bellied flycatchers seen at Blakeney Point
Yellow-bellied blow-ins rest at Blakeney Point

Tower block falcons fall
Tate Modern falcons, falcons on St Paul's

Buzzards gather, circle overhead
Buzzards on the interstate, something must be dead

Red-winged blackbirds tumble down, down in Arkansas
Red-winged blackbirds on the ground, down in Arkansas

The cuckoo she's a pretty bird, she warbles as she flies
The cuckoo was a pretty bird, she never said goodbye

NOTES

A *plastiglomerate* is a rock that forms when plastic is melded with shells, sand and other sedimentary material by fire. The phenomenon was first identified and named in 2006 by the oceanographer Charles Moore on Kamilo Beach, Hawai'i.

THE TWO MAGICIANS

The first three stanzas are from the traditional folk ballad 'The Twa Magicians' as collected by Francis James Child and published in his English and Scottish Popular Ballads between 1882 and 1898. The list of chemical names are alternative trade names for dichlorodiphenyltrichloroethane or DDT. *The Act for the better Cultivation...* is also called *The Enclosure Act of 1773.* The act enabled the enclosure of land and removed the right of common access. The list on page 29 is the list of objects found in the stomach of a dead 36-foot grey whale found on a beach near Seattle in May 2010. The reference to the 'circling sky' is from John Clare's poem 'The Mores' (1821-1831). The lists towards the end of the poem include names of winds, forms of gas, coal and oil extracted for energy, and oil fields in the North Sea and in Alberta Canada (Abathasca). The quotes in italics are from Bruce Springsteen's 'Born to Run' and Pasty Cline's 'Crazy'. Harvey and Irma were the names given to hurricanes that devastated parts of the Caribbean and United States in the summer of 2017.

BEACHED

The references to #explodingwhale and hasthewhaleexplodedyet. com are from social media around the appearance of a dead blue whale on the coast of Newfoundland in May 2014.

FRIENDLY FLOATEES/TRIPADVISOR

This poem is constructed by looking up the beaches where the small plastic bath toys were known to have been found between 1992 and 2006 on Tripadvisor and using language found in the reviews for these beaches.

FOOTFALL

This poem is heavily indebted to Jean Sprackland's poem 'Breaking the Fall' from her collection *Tilt* (2007).

BLUES FOR LOST BIRDS

The last couplet is from the popular folk song 'The Cuckoo'. The number of cuckoos in the UK has dropped by 65% since the 1980s.